ENERGY SECTOR STANDARD OF THE PEOPLE'S REPUBLIC OF CHINA

中华人民共和国能源行业标准

Code for Design of Densified Biofuel Boiler Plants

生物质成型燃料锅炉房设计规范

NB/T 10240-2019

Chief Development Department: China Renewable Energy Engineering Institute
Approval Department: National Energy Administration of the People's Republic of China
Implementation Date: May 1, 2020

China Water & Power Press

中国水利水电出版社

Beijing 2024

All rights reserved. No part of this publication may be reproduced, stored in a retrieval system, or transmitted in any form or by any means—electronic, mechanical, photocopying, recording or otherwise, without prior written permission of the publisher.

图书在版编目（CIP）数据

生物质成型燃料锅炉房设计规范 : NB/T 10240-2019 = Code for Design of Densified Biofuel Boiler Plants (NB/T 10240-2019) : 英文 / 国家能源局发布. 北京 : 中国水利水电出版社, 2024. 10. -- ISBN 978-7-5226-2797-7

Ⅰ. TK22-65

中国国家版本馆CIP数据核字第2024KM3895号

ENERGY SECTOR STANDARD
OF THE PEOPLE'S REPUBLIC OF CHINA
中华人民共和国能源行业标准

Code for Design of Densified Biofuel Boiler Plants
生物质成型燃料锅炉房设计规范
NB/T 10240-2019
（英文版）

Issued by National Energy Administration of the People's Republic of China
国家能源局　发布
Translation organized by China Renewable Energy Engineering Institute
水电水利规划设计总院　组织翻译
Published by China Water & Power Press
中国水利水电出版社　出版发行
　　Tel: (+ 86 10) 68545888　68545874
　　sales@mwr.gov.cn
　　Account name: China Water & Power Press
　　Address: No.1, Yuyuantan Nanlu, Haidian District, Beijing 100038, China
　　http://www.waterpub.com.cn
中国水利水电出版社微机排版中心　排版
北京中献拓方科技发展有限公司　印刷
184mm×260mm　16开本　3.5印张　111千字
2024年10月第1版　2024年10月第1次印刷
Price（定价）：￥450.00

Introduction

This English version is one of China's energy sector standard series in English. Its translation was organized by China Renewable Energy Engineering Institute authorized by National Energy Administration of the People's Republic of China in compliance with relevant procedures and stipulations. This English version was issued by National Energy Administration of the People's Republic of China in Announcement [2023] No. 8 dated December 28, 2023.

This version was translated from the Chinese Standard NB/T 10240-2019, *Code for Design of Densified Biofuel Boiler Plants*, published by China Water & Power Press. The copyright is reserved by National Energy Administration of the People's Republic of China. In the event of any discrepancy in the implementation, the Chinese version shall prevail.

Many thanks go to the staff from the relevant standard development organizations and those who have provided generous assistance in the translation and review process.

For further improvement of the English version, any comments and suggestions are welcome and should be addressed to:

China Renewable Energy Engineering Institute
No. 2 Beixiaojie, Liupukang, Xicheng District, Beijing 100120, China
Website: www.creei.cn

Translating organizations:

POWERCHINA Northwest Engineering Corporation Limited

China Renewable Energy Engineering Institute

Translating staff:

| GAO Xujun | MA Bo | LI Zhongjie | LEI Yongzhi |
| CHEN Wei | QIAO Peng | | |

Review panel members:

GUO Jie	POWERCHINA Beijing Engineering Corporation Limited
QIE Chunsheng	Senior English Translator
YAN Wenjun	Army Academy of Armored Forces, PLA
YE Bin	POWERCHINA Huadong Engineering Corporation Limited

CHEN Lei	POWERCHINA Zhongnan Engineering Corporation Limited
FANG Shunli	Xi'an Thermal Power Research Institute Corporation Limited
LIU Yanfeng	Xi'an University of Architecture and Technology
YUE Lei	China Renewable Energy Engineering Institute
LI Shisheng	China Renewable Energy Engineering Institute

National Energy Administration of the People's Republic of China

翻译出版说明

本译本为国家能源局委托水电水利规划设计总院按照有关程序和规定，统一组织翻译的能源行业标准英文版系列译本之一。2023年12月28日，国家能源局以2023年第8号公告予以公布。

本译本是根据中国水利水电出版社出版的《生物质成型燃料锅炉房设计规范》NB/T 10240—2019翻译的，著作权归国家能源局所有。在使用过程中，如出现异议，以中文版为准。

本译本在翻译和审核过程中，本标准编制单位及编制组有关成员给予了积极协助。

为不断提高本译本的质量，欢迎使用者提出意见和建议，并反馈给水电水利规划设计总院。

地址：北京市西城区六铺炕北小街2号
邮编：100120
网址：www.creei.cn

本译本翻译单位：中国电建集团西北勘测设计研究院有限公司
　　　　　　　　水电水利规划设计总院

本译本翻译人员：高徐军　马　勃　李仲杰　雷永智
　　　　　　　　陈　伟　乔　鹏

本译本审核人员：

　　郭　洁　中国电建集团北京勘测设计研究院有限公司
　　郄春生　英语高级翻译
　　闫文军　中国人民解放军陆军装甲兵学院
　　叶　彬　中国电建集团华东勘测设计研究院有限公司
　　陈　蕾　中国电建集团中南勘测设计研究院有限公司
　　方顺利　西安热工研究院有限公司
　　刘艳峰　西安建筑科技大学
　　岳　蕾　水电水利规划设计总院
　　李仕胜　水电水利规划设计总院

国家能源局

Announcement of National Energy Administration of the People's Republic of China
[2019] No. 6

National Energy Administration of the People's Republic of China has approved and issued 384 energy sector standards including *Technical Specification for Electrical Exploration of Hydropower Projects* (Attachment 1), the English version of 48 energy sector standards including *Technical Guide for Rock-Filled Concrete Dams* (Attachment 2), and Amendment Notification No. 1 for 7 energy sector standards including *Technical Code for Environmental Impact Assessment of Wind Farm Projects* (Attachment 3); and abolished 5 energy sector standards/plans including *Charging Standards for Investigation and Design of Wind Power Project* (Attachment 4).

Attachments: 1. Directory of Sector Standards
2. Directory of English Version of Sector Standards
3. Amendment Notification for Sector Standards
4. Directory of Abolished Sector Standards/Plans

National Energy Administration of the People's Republic of China

November 4, 2019

Attachment 1:

Directory of Sector Standards

Serial number	Standard No.	Title	Replaced standard No.	Adopted international standard No.	Approval date	Implementation date
...						
17	NB/T 10240-2019	Code for Design of Densified Biofuel Boiler Plants			2019-11-04	2020-05-01
...						

Foreword

According to the requirements of Document GNKJ [2015] No. 12 issued by National Energy Administration of the People's Republic of China, "Notice on Releasing the Development and Revision Plan of the Second Batch of Energy Sector Standards in 2014", and after extensive investigation and research, summarization of practical experience, and wide solicitation of opinions, the drafting group has prepared this code.

The main technical contents of this code include: boiler plant layout; handling and storage of fuel; fuel conveying system; boiler; boiler flue gas and air system; ash and slag removal system; boiler feedwater; thermotechnical monitoring and control; chemical test; civil works, electrical and utility system; environmental protection and energy saving; fire protection; occupational health and safety; operation and maintenance.

National Energy Administration of the People's Republic of China is in charge of the administration of this code. China Renewable Energy Engineering Institute has proposed this code, and is responsible for its routine management and the explanation of specific technical contents. Comments and suggestions in the implementation of this code should be addressed to:

China Renewable Energy Engineering Institute
No. 2 Beixiaojie, Liupukang, Xicheng District, Beijing 100120, China

Chief development organizations:

POWERCHINA Northwest Engineering Corporation Limited

China Renewable Energy Engineering Institute

Participating development organizations:

HONGRI (Great Resources)

Devotion Corporation

Academy of Agricultural Planning and Engineering, MARA

Beijing Aoke Ruifeng New Energy Co., Ltd.

Xi'an University of Architecture and Technology

Chief drafting staff:

ZHANG Peng	ZHOU Zhi	HONG Hao	ZHANG Maoyong
RUAN Liangwei	HU Xiaofeng	WANG Zhenkun	XU Chenchen
WANG Jilin	SUI Hairan	CHEN Wei	WANG Biao

| MENG Haibo | ZHAO Lixin | YAO Zonglu | LI Angui |

Review panel members:

XIE Hongwen	YANG Zhigang	YU Jiaqing	ZHAO Ming
LI Youjin	CHEN Ping	LI Dingkai	TIAN Gang
YU Guosheng	TIAN Jingkui	TIAN Xinmin	YAN Zhanliang
ZHANG Jian	YANG Zhiming	WANG Chao	ZHOU Caiquan
LI Shisheng			

Contents

1	**General Provisions**	1
2	**Terms**	2
3	**Basic Requirements**	3
4	**Boiler Plant Layout**	4
4.1	Site Selection	4
4.2	General Layout	4
4.3	Arrangement of Buildings and Structures	6
4.4	Process Setup	6
5	**Handling and Storage of Fuel**	8
5.1	Fuel Handling	8
5.2	Fuel Storage	8
6	**Fuel Conveying System**	10
6.1	Charging	10
6.2	Conveying	11
6.3	Feeding	12
7	**Boiler**	13
7.1	Boiler Equipment	13
7.2	Boiler Auxiliary System	14
8	**Boiler Flue Gas and Air System**	15
8.1	Flue Gas and Air System	15
8.2	Purification and Emission of Flue Gas	15
9	**Ash and Slag Removal System**	18
10	**Boiler Feedwater**	19
11	**Thermotechnical Monitoring and Control**	20
11.1	Thermotechnical Monitoring	20
11.2	Control	20
12	**Chemical Test**	24
13	**Civil Works, Electrical and Utility System**	25
13.1	Civil Works	25
13.2	Electrical	25
13.3	Heating, Ventilation and Air Conditioning	26
13.4	Water Supply and Drainage	26
14	**Environmental Protection and Energy Saving**	28
14.1	Pollution Prevention and Control	28
14.2	Greening	30
14.3	Environmental Management and Monitoring	30

14.4	Energy Saving	31
15	**Fire Protection**	**32**
16	**Occupational Health and Safety**	**35**
16.1	Occupational Safety	35
16.2	Occupational Health	36
17	**Operation and Maintenance**	**37**
17.1	Boiler Operation	37
17.2	Maintenance	37
17.3	Emergency Response Plan	38

Explanation of Wording in This Code **39**
List of Quoted Standards **40**

1 General Provisions

1.0.1 This code is formulated with a view to standardizing the design of densified biofuel boiler plants.

1.0.2 This code is applicable to the design of the construction, renovation, and extension of the following densified biofuel boiler plants:

1. Steam boiler plants with water as the medium. The rated evaporation for a single boiler is 10 t/h to 65 t/h and the rated steam pressure is 1.0 MPa to 2.5 MPa.

2. Hot water boiler plants. The rated thermal power for a single boiler is 7 MW or above and the rated discharge pressure is 0.7 MPa to 2.5 MPa.

3. Organic fluid boiler plants. The rated thermal power for a single boiler is 7 MW to 20 MW.

1.0.3 In addition to this code, the design of densified biofuel boiler plants shall comply with other current relevant standards of China.

2 Terms

2.0.1 densified biofuel heating plant

heating complex using densified biofuel boiler plant as the heat source

2.0.2 district boiler plant

boiler plant that serves a certain area where there might be a number of enterprises, civil and public buildings, and other building facilities

3 Basic Requirements

3.0.1 The design of densified biofuel boiler plants shall make reasonable use of local resources and focus on the near-term planning while considering the medium- to long-term planning according to the overall district or enterprise planning and the district heat supply planning, and should leave room for renovation and extension.

3.0.2 Reliable and sustainable supply of densified biofuel should be available in the district where the boiler plant is located.

3.0.3 The preliminary investigation for boiler plant design shall determine the types and sources of densified biofuel and obtain the corresponding fuel supply agreement and related basic data such as biofuel element analysis and industrial analysis results, meteorology, geology, hydrology, electricity and water supply.

3.0.4 The boiler plant design shall take effective measures to mitigate the environmental impact of waste gas, waste water, waste residue, and noise, to ensure that the pollutant emission and noise indicators comply with the provisions of the current relevant standards of China.

3.0.5 For renovation and extension of densified biofuel boiler plants, the design shall obtain the information on the original process equipment and pipelines, and shall be conducted in a holistic and coordinated manner with due consideration of the original general plan layout, process system, building structure, and operational experience, etc.

3.0.6 For the densified biofuel boiler plants built in regions with a seismic intensity of Ⅵ or above, a seismic design shall be performed for the buildings, structures, equipment, and pipelines.

4 Boiler Plant Layout

4.1 Site Selection

4.1.1 The boiler plant site shall be determined through comprehensive techno-economic evaluation according to the overall district or enterprise planning, heat supply planning, resources distribution, and the densified biofuel supply capacity, with due consideration of the natural environment, construction conditions, and socioeconomic factors for site selection.

4.1.2 The boiler plant site selection shall meet the following requirements:

1. Close to the areas with concentrated heating load, and technically feasible and cost-effective for the arrangement of heating pipelines and outdoor heating pipe network.

2. Conducive to densified biofuel storage and ash and slag discharge, and preferably separating material flow from human flow.

3. There should be land reserve for planned extension.

4. Conducive to natural ventilation and lighting.

5. Beneficial to mitigating the impact of air pollutants, noise, and ash and slag on environmentally sensitive targets. The boiler plants that operate all year round shall be located at the upwind side of the minimum annual frequency wind direction in the target area, and the boiler plants that operate seasonally shall be located at the downwind side of the maximum frequency wind direction in the season.

6. In addition to the above-mentioned requirements, the location of the boiler plant for an inflammables and explosives producer shall also comply with the provisions of the relevant sector codes.

4.1.3 The overall planning of the boiler plant shall be coordinated with the planning and layout of the district heating pipeline.

4.2 General Layout

4.2.1 The general layout planning of the boiler plant shall be made in a holistic manner according to the process flow, transportation, fire protection and prevention, explosion prevention, environmental protection, safety and health, construction and living requirements, etc., taking into account the natural conditions such as topography, geology, earthquakes and meteorology as well as heating load. The general layout shall center around the boiler house and be compact and reasonable with clear functional zoning, to streamline the process flow and facilitate maintenance and repair, construction, and extension.

4.2.2 The general layout planning for the extension or renovation works in the boiler plant shall consider the existing production system and its layout, reasonably make use of existing facilities to reduce demolition and relocation, and mitigate the impact of extension and renovation on the existing production system.

4.2.3 Under the premise of meeting the process requirements, the orientation and location of each building and structure of the boiler plant should be determined taking into account the location conditions, sunlight, natural ventilation and lighting, etc.

4.2.4 The vertical layout in the densified biofuel boiler plant area shall consider the process requirements, engineering geology, hydro-meteorology, earth and rock work volume and foundation treatment, and shall meet the following requirements:

  1 The elevations of buildings, structures, roads, etc. shall be determined such that facilitates production and maintenance. The elevations and arrangement for the foundations, pipelines, pipe racks, pipe trenches in the ground and underground facilities, and the basements shall be planned in a unified manner to achieve reasonable crossover, convenient maintenance and extension, and smooth drainage.

  2 Cut-fill balance shall be achieved in the plant area and construction site. When impossible, the borrow area and disposal area shall be determined.

  3 The plant area shall have such a minimum slope and aspect that no water stands on the ground, which shall be compatible with the arrangement of the rainwater cellar wells and rainwater outlets of buildings, structures, roads and sites, and shall be determined according to factors such as local rainfall and site soil conditions.

4.2.5 Vehicle lanes, fire lanes and sidewalks shall be arranged between the buildings and structures in the boiler plant according to the needs of production, living and firefighting. The road design shall comply with the current national standard GB 50187, *Code for Design of General Layout of Industrial Enterprises*.

4.2.6 The number of the main entrances to densified biofuel heating plant area should not be less than 2, and main pedestrian entrances should be separated from the main material flow entrances.

4.2.7 Flood control and waterlogging prevention facilities of the boiler plant shall meet the flood control standards of the surrounding area.

4.3 Arrangement of Buildings and Structures

4.3.1 The plan layout and spatial combination of buildings and structures in the boiler plant area shall have clear functional zoning, be compact and reasonable, concise and harmonized in architecture, to enable streamlined process flow, operational safety, and easy transportation, installation, and maintenance.

4.3.2 The spacing between the boiler house, densified biofuel storehouse, dry slag room, dry ash storage bin and chimney, as well as their clearance to other buildings and structures shall comply with the current national standard GB 50016, *Code for Fire Protection Design of Buildings*, and meet the requirements for installation, operation and maintenance.

4.3.3 The layout of buildings and structures for the densified biofuel conveying system shall meet production process requirements, and shall take advantage of terrain to shorten the conveying distance, reduce the transfer, and lower the lifting height.

4.3.4 The densified biofuel storehouse and dry ash storage bin should be arranged close to the material transportation entrance and exit, respectively, and shall be located on the upwind side of the annual minimum frequency wind direction in the densified biofuel heating plant area.

4.4 Process Setup

4.4.1 The process setup of the boiler plant shall comply with the heat production process and guarantee the safety and ease of equipment installation, operation, and maintenance, and shall achieve a reasonable equipment layout, convenient and neat connection of flue gas and air ducts and water and vapor pipelines, compact space, and less land occupation.

4.4.2 When a boiler is installed in the open air, the following requirements shall be met:

1 The boiler proper and its auxiliaries suitable for outdoor conditions shall be selected.

2 Measurement and control instruments for water level, pressure, temperature and flow shall be set in an instrument control room. For measurement and control instruments and pipeline valves and other accessories, measures shall be taken against rain and snow, wind, frost, corrosion, and heat loss.

3   Fans, water tanks, deaerators, heating devices, dust removers, heat accumulators, water treatment devices and other auxiliaries shall be provided with measures against rain and snow, wind, frost, corrosion, and noise.

4.4.3   The layout of main valves and their servo actuators and thermotechnical monitoring and control devices, etc., shall facilitate their operation, maintenance and repair, and safety platforms and ladders shall be installed when necessary.

4.4.4   The clear height of the operation site and passage shall not be less than 2 m and shall allow for the operation of lifting equipment.

5 Handling and Storage of Fuel

5.1 Fuel Handling

5.1.1 Densified biofuel may be directly unloaded into the storehouse by forklift, crane lifting, or manual handling, and may also be unloaded mechanically by truck unloading trough, bottom-dumping bunker, and bulk material handling.

5.1.2 For materials packaged in ton bags, forklift, crane lifting, etc. may be employed for unloading, and the following requirements shall be met:

1 The crane lifting capacity may be designed to be 2 times to 3 times the rated load.

2 The support frame shall be installed at the discharge port to prevent ton bags from falling.

5.1.3 When a truck unloading trough is used, the discharging capacity of the unloading trough shall match the unloading capacity.

5.1.4 When a bottom-dumping bunker is used, its discharging capacity shall match the unloading capacity and system conveying capacity.

5.1.5 Bulk materials should be unloaded directly from the dump truck to the material pit, and the hopper throat shall be properly connected with the dump truck.

5.1.6 The unloading lift pit shall be provided with measures to prevent leaks and water seepage; a canopy shall be installed on the top of the lift to prevent rainwater from entering the lift and the pit.

5.2 Fuel Storage

5.2.1 Measures shall be taken for the storage of densified biofuel against rain, water, and spontaneous combustion. Enclosed or semi-enclosed densified biofuel storehouse should be provided.

5.2.2 The design storage capacity of the storehouse shall not be less than 5-d biofuel consumption of the boiler. The storage life of the densified biofuel should not exceed 180 d.

5.2.3 The design of densified biofuel storehouse shall consider the following factors:

1 The length, width, and span of the storehouse shall be determined according to the general plan layout of the boiler plant, the storage life, the dimensions of the biofuel package, the number of the biofuel

packages taken by unloading and grabbing devices at one time, etc.

2   The height of the storehouse shall be determined according to the erection height of the unloading and grabbing devices, the maximum height during device operation, the height of the stored biofuel packages, etc.

3   The width and height of the storehouse entrance shall be determined according to the maximum dimensions of transport vehicles at full load.

4   The storehouse shall be equipped with facilities for ventilation, dedusting, anti-explosion, lighting, firefighting, etc., and they shall be easy to clean.

5.2.4   When stockpiling the materials in ton bags or fractional packs in densified biofuel storehouse, the following requirements shall be met:

1   Materials packaged in ton bags shall be stacked neatly on pallets, around which operation channels shall be reserved for auxiliary operation machinery, to allow monorail cranes or forklifts to load materials.

2   The stacking height of materials packaged in fractional packs shall not exceed 2.5 m, and the side of the stockpile should be in a regular trapezoid shape and shall be at least 0.5 m away from the channel.

5.2.5   The design of silos shall meet the following requirements:

1   The silo bottom shall have an inclination of not less than 60° to prevent dead corner in stacking. Enough manholes shall be provided at the upper, middle and lower parts for regular cleaning of the silo.

2   Besides the material level monitoring system, a carbon monoxide detection device shall be installed inside the silo; air rapping apparatus may be installed at the cone bucket.

3   The air breathing valve and explosion vent shall be installed on the silo roof for releasing the combustible gas and pressure in case of accident.

6 Fuel Conveying System

6.1 Charging

6.1.1 The design of conveying and charging system shall meet the requirements of streamlined equipment, short path, efficient conveying, low crushing rate and less dust; the space and platform for repair and maintenance shall be reserved.

6.1.2 Two independent charging systems should be designed for densified biofuel conveying, one in service and one on standby. The capacity of each charging system shall be 1.5 times the design biofuel consumption per unit time.

6.1.3 Depending on the types of densified biofuel, the conveying and charging system may employ bucket lifts, belt conveyors, flight conveyors, screw conveyors, etc.

6.1.4 When the densified biofuel is stored in the storehouse, the charging approaches shall be designed according to package forms, and the following requirements should be met:

   1   For materials packaged in ton bags, forklifts, monorail cranes, cantilever cranes, etc. should be used to carry the fuel to the charging inlet, and bucket lifts, belt conveyors, flight conveyors, and screw conveyors should be used to convey the fuel to the bunker at the boiler.

   2   For materials packaged in fractional packs, bucket lifts, belt conveyors, flight conveyors, etc. should be used to convey the fuel from the charging inlet to the bunker at the boiler.

   3   For bulk materials, forklifts should be used to carry the fuel to the charging inlet, and bucket lifts, belt conveyors, flight conveyors, etc. should be used to convey the fuel to the bunker at the boiler.

6.1.5 When densified biofuel is stored in silos, a screw collector should be set at the bottom of the silo, and hoppers, screw conveyors, belt conveyors, bucket lifts, etc. should be used to convey the fuel to the bunker at the boiler.

6.1.6 The design capacity of the bunker at the boiler shall not be less than 1-h fuel consumption of the boiler operating at the rated heating load.

6.1.7 Densified biofuel storehouses, silos, and conveying systems should be arranged on the side of the bunker at the boiler.

6.1.8 The conveying and charging system equipment shall be equipped with a gang switch supporting "clockwise-on and counterclockwise-off".

6.2 Conveying

6.2.1 The bucket lift system for densified biofuel shall meet the following requirements:

1. For steam boilers with a rated evaporation capacity of 20 t/h or below and for hot water boilers with a rated thermal power of 14 MW or below, a bucket lift may be used for the charging and conveying system. The bucket lift may be arranged at the front end or left or right side of the hopper depending on the densified biofuel storehouse layout and the feed-in direction of the boiler.

2. In the case of one bucket lift, it shall be arranged at the front end or on one side of the hopper. In the case of two bucket lifts, they shall be arranged respectively on the left and right sides of the hopper. The lifting angle of the bucket lift shall be 90°.

3. A movable access port shall be set at the bucket lift bottom for easy inspection and maintenance at any time.

6.2.2 The flight conveyor for densified biofuel shall meet the following requirements:

1. When fuel conveying has an angle requirement, a flight conveyor may be used. The flight conveyor may be arranged at the front end or on one side of the hopper, and the conveying angle should not be greater than 30°.

2. The conveying distance shall be limited within 10 m, and an outlet shall be set at the end of the flights.

6.2.3 The belt conveying system for densified biofuel shall meet the following requirements:

1. When fuel is to be conveyed in a large amount and at a small angle, a belt conveyor may be used. The belt conveyor may be arranged at the front end or on one side of the hopper.

2. When a flat belt conveyor is used, its conveying angle should not be greater than 16°.

3. When a skirt baffle belt conveyor is used, its conveying angle should not be greater than 45°.

4. The belt conveyor shall be equipped with an easy-to-remove bottom plate to facilitate cleaning the bottom.

6.2.4 When the requirement for particle size of densified biofuel is high and the conveying volume is small, a screw conveyor may be used. The screw conveyor shall be arranged horizontally at the front end of the hopper. The number of screw conveyors shall be determined according to the boiler fuel consumption and the screw conveyor output, and should be 2 to 4.

6.3 Feeding

6.3.1 Feeding methods shall be selected according to the size and physical properties of densified biofuel. Screw feeding, belt feeding, star feeding, and free-fall feeding may be adopted.

6.3.2 Fluidized-bed boilers should adopt a screw feeder, and grate-fired boilers may adopt a star feeder or a free-fall feeder. The number of feeders shall be determined according to the boiler capacity and feeder output.

6.3.3 For various densified biofuel, the screw feeding shall meet the following requirements:

1 For granular densified biofuel, a small screw feeder may be used for longitudinal feeding. To ensure uniform feeding, multiple feeders may be used simultaneously according to the furnace capacity.

2 For blocky densified biofuel, a shaftless screw feeder may be used for radial feeding, and a space shall be reserved for handling the jam.

3 The screw feeder shall be reversible for clearing the jam.

6.3.4 In the case of a bunker at the boiler, there shall be a space for removing biofuel from the bunker; when a screw feeder is employed at the bunker bottom, a space shall be reserved for disassembling and assembling the screw shaft.

6.3.5 The boiler feed port shall be equipped with real-time monitoring equipment.

6.3.6 The boiler feeding system shall meet the following requirements:

1 The automatic bunker material level control system shall be provided, with sound and light alarm functions for low material levels.

2 The bunker temperature monitoring system shall be provided, with sound and light alarm functions for over temperature.

3 The bunker smoke detection and sound and light alarm function shall be provided.

7 Boiler

7.1 Boiler Equipment

7.1.1 The main technical parameters of the boiler shall comply with the current sector standard NB/T 47062, *Biomass Molded Fuel Fired Boilers*.

7.1.2 Boiler type selection should meet the following requirements:

1. When the rated evaporation capacity of a steam boiler is greater than 20 t/h and the fluctuation of steam consumption is less than or equal to 30 %, or the rated thermal power of a hot water boiler is greater than 14 MW, a circulating fluidized bed boiler should be used.

2. When the rated evaporation capacity of a steam boiler is less than or equal to 20 t/h and the fluctuation of steam consumption is greater than 30 %, or the rated thermal power of a hot water boiler is less than or equal to 14 MW, a grate-fired boiler should be used.

3. When the load fluctuation frequency of a grate-fired boiler is 0.5 t/min, a traveling grate should be used; in the case of a large load fluctuation frequency, an inclined reciprocating grate should be used.

7.1.3 The design capacity of the boiler plant and the number of boilers shall meet the following requirements:

1. The design capacity of the boiler plant shall be determined according to the comprehensive maximum heating load of the heating system.

2. The number of boilers in the boiler plant should not be less than 2; however, a single boiler may be installed only if it can meet the heating load and maintenance needs.

3. The design capacity of a single boiler shall ensure a high operating efficiency for a long time, and the actual operating load rate should not be lower than 50 %.

4. When there are multiple boilers in the boiler plant, the capacity of each boiler should be the same. Otherwise, when the boiler with the highest rated evaporation capacity or thermal power is under overhaul, the remaining boilers shall meet the following requirements:

    1) The minimum heating load required for continuous production shall be met.

    2) The minimum heating load required for heating, ventilation and air conditioning and domestic heat shall be met. The total heat supply

of the remaining boilers shall not be lower than 65 % of the design heat supply for heating in cold region and not be lower than 70 % for severe cold region.

7.2 Boiler Auxiliary System

7.2.1 Grate-fired boilers may use automatically controlled hot air ignition with electric heating or automatically controlled ignition with auxiliary fuels; fluidized bed boilers may use under-bed ignition or bed ignition, and under-bed ignition should be given priority.

7.2.2 The primary and secondary air systems of the boiler shall be installed independently. The air supply may be adjustable according to the nature of densified biofuel and load changes to make the corresponding air volume and pressure meet the stable operation requirements of the boiler.

7.2.3 Boiler circulating water pump shall be adjustable to the change of boiler load to make the corresponding flow and lift meet the stable operation requirements of the equipment.

7.2.4 The fans and pumps should be equipped with variable-frequency or energy-efficient motors, whose noise and vibration control shall comply with the current national standard GB 50041, *Code for Design of Boiler Plant*.

8 Boiler Flue Gas and Air System

8.1 Flue Gas and Air System

8.1.1 The boiler air system shall have good airtightness, and the air distribution shall be flexible and effective. Under rated conditions, the excess air coefficient α at the smoke exhaust of the last heating surface of the boiler shall meet the following requirements:

1. For grate-fired boilers, the α of boilers with stoker-fired grate shall not be higher than 1.6; the α of boilers with fixed grate shall not be higher than 1.65.

2. The α of fluidized bed boilers and boilers with membrane walls shall not be higher than 1.4.

8.1.2 The boiler shall be equipped with a secondary air system. According to the fuel characteristics and furnace structure, the position and number of secondary air nozzles shall be such that the densified biofuel, volatile combustible gases, and air are fully mixed and burned. Moreover, the secondary air branch pipe shall be provided with a damper that can regulate the air volume separately.

8.1.3 Under the maximum boiler load, the induced draft fan shall enable the negative pressure operation of the furnace at −30 Pa to −50 Pa.

8.1.4 The air volume and pressure of blowers and induced draft fans shall enable the stable operation of the boiler at the rated output. The excess air volume should not be less than 10 % of the calculated air volume, and the excess air pressure should not be less than 20 % of the calculated air pressure; the air volume and pressure shall have required adjustment range and flexibility.

8.2 Purification and Emission of Flue Gas

8.2.1 The boiler plant shall be equipped with purification and soot removal devices, and adopt low NO_x combustion technology, and when necessary, shall be provided with denitrification and desulfurization facilities, to make the flue gas emission meet the special requirements for air pollutants from gas-fired boilers in the current national standard GB 13271, *Emission Standard of Air Pollutants for Boiler*.

8.2.2 The measured emission concentrations of particulate matter, sulfur dioxide, and nitrogen oxides shall be converted to the emission concentration at baseline oxygen content. The baseline oxygen content of the boiler shall be

9 % as specified in the current national standard GB 13271, *Emission Standard of Air Pollutants for Boiler*.

8.2.3 The boiler may adopt the two-stage soot removal. A cyclone dust collector should be selected for the primary soot removal and a bag dust collector should be selected for the secondary soot removal. The total soot removal efficiency shall reach 99.90 % at least, and the soot removal devices can operate stably for a long time.

8.2.4 The air compressor shall have the capacity to provide enough compressed air for the baghouse blowing and soot blowing, and other uses, leaving a certain margin.

8.2.5 Selective non-catalytic reduction (SNCR) shall be given priority for flue gas denitrification, while selective catalytic reduction (SCR) may be used according to the actual situations.

8.2.6 The arrangement of the boiler plant chimney shall meet the following requirements:

1. Only one chimney is allowed to be set for each new boiler plant. The lining of the chimney should be designed to emit weakly corrosive flue gas.

2. The chimney height shall comply with the current national standard GB 13271, *Emission Standard of Air Pollutants for Boiler*, and shall not be lower than 8 m. The specific height shall be determined in accordance with the approved environmental impact assessment documents.

3. Where there are buildings within 200 m around the chimney of a new boiler plant, the chimney shall be 3 m higher than the highest building.

4. Sampling of exhaust gas from the boiler shall be conducted at the specified pollutant emissions monitoring points according to the type of pollutants monitored. Where there are exhaust gas treatment facilities, monitoring shall be conducted behind the facilities. Monitoring and sampling locations and setting for air pollutants in exhaust gas shall comply with the current standards of China GB 5468, *Measurement Method of Smoke and Dust Emission from Boilers*; GB/T 16157, *The Determination of Particulates and Sampling Methods of Gaseous Pollutants Emitted from Exhaust Gas of Stationary Source*; and HJ/T 397, *Technical Specifications for Emission Monitoring of Stationary Source*.

5. The setting of the online monitoring system for flue gas shall comply

with the current sector standards HJ/T 397, *Technical Specifications for Emission Monitoring of Stationary Source*; and HJ 75, *Specifications for Continuous Emissions Monitoring of SO_2, NO_X, and Particulate Matter in the Flue Gas Emitted from Stationary Sources*.

9 Ash and Slag Removal System

9.0.1 The ash and slag removal system shall be selected according to the type of slag eliminator and soot blower, ash and slag amount and properties, haul distance, site landform, meteorological conditions, transportation conditions, environmental protection, comprehensive utilization, etc. The ash and slag should be removed separately.

9.0.2 All ash and slag shall be utilized. An ash and slag room shall be set on site, with a storage capacity no less than the calculated 7-d maximum ash and slag output of the boiler plant.

9.0.3 The slagging machine shall be of closed type.

9.0.4 The ash removal system may adopt such types as the sealed bag, airtight pneumatic conveying, or shared ash and slag removal system according to ash removal method, ash volume, land occupation of the project, etc.

9.0.5 The ash from the soot blower and economizer may be transported to a dry ash storage bin by an airtight pneumatic conveying system, and the following requirements shall be met:

1 The storage capacity of the dry ash storage bin shall be the calculated 2-d to 3-d maximum ash output of the boiler plant.

2 The ash conveying pump shall be equipped with pressure and material level indicators. The ash pipe elbow shall be of a large-radius prefabricated type and lined with wear-resistant material.

3 The dry ash storage bin shall have medium- and high-level indicators. When the dry ash storage volume reaches medium or high levels, the dry ash storage bin shall be able to ensure the normal operation of the boiler plant system for at least 12 h or 2 h, respectively.

4 The dry ash storage bin shall be equipped with fluidizing devices with flat and smooth surfaces.

5 The dry ash storage bin shall be provided with a dry powder bulk loader at the discharge outlet, and the discharge position and height of the bulk loader shall well match the powder inlet of the tanker.

10 Boiler Feedwater

10.0.1 The boiler feedwater system shall meet the following requirements:

1 The boiler may adopt one route of feedwater.

2 If the interruption of feedwater might cause significant losses, two routes of feedwater taken from different pipe sections of outdoor ring network or different water sources shall be adopted.

3 When adopting one route of feedwater, a water tank or pool shall be set up, and the total capacity shall not be less than the sum of the capacities of the original water tank, softened or demineralized water tank, deaerated water tank, and intermediate water tank, and shall not be less than the calculated 2-h water consumption of the boiler plant.

10.0.2 The quality of boiler feedwater, boiler water, makeup water, and circulating water shall comply with the current national standard GB 1576, *Water Quality for Industrial Boilers*. The water quality shall also meet the special requirements indicated in the boiler user's manual.

11 Thermotechnical Monitoring and Control

11.1 Thermotechnical Monitoring

11.1.1 The setup of monitoring instruments and alarm signal devices for steam and hot-water boilers shall comply with the current standards of China GB 50041, *Code for Design of Boiler Plant*; and TSG G0001, *Boiler Safety Technical Supervision Administration Regulation*.

11.1.2 The safe and economic operation parameters monitoring instruments installed in the organic fluid boiler plants shall comply with current sector standard SY/T 0524, *Code for Heater System with Heat-Transfer Oil*.

11.2 Control

11.2.1 The boiler control system shall comply with the current sector standards TSG G0001, *Boiler Safety Technical Supervision Administration Regulation*; TSG G0002, *Supervision Administration Regulation on Energy Conservation Technology for Boiler*; NB/T 47034, *Specification for Industrial Boilers*; and NB/T 47051, *Technical Specification for Industrial Boiler Control Devices*.

11.2.2 The control system shall be selected with consideration of boiler capacity, complexity and reliability of operation control, the construction investment, the convenience of operation, etc. When the rated evaporation capacity is greater than 20 t/h or the rated thermal power is greater than 14 MW, a distributed control system (DCS) may be used. When the rated evaporation capacity is less than or equal to 20 t/h or the rated thermal power is less than or equal to 14 MW, a programmable logic controller (PLC) may be used.

11.2.3 The boiler control system shall be provided with sound or sound-light alarms which is able to distinguish different parameters, including over-limit alarms for pressure, temperature, water level, material level, and other safe operation parameters. Upon alarm, the boiler control system shall be able to immediately initiate the interlock protection measures.

11.2.4 In the case of failure of sensors or their circuits, the boiler control system shall enable alarm and interlock protection.

11.2.5 Electrical interlocking devices shall be provided between the equipment of densified biofuel conveyance system. Shutdown in sequence shall be available in normal operation, and the delay time shall allow for no-load restart.

11.2.6 The boiler control system shall be provided with electrical interlock, which shall automatically cut off the boiler feed when the combustion chamber temperature is lower than the specified value or when the furnace chamber pressure exceeds the limit.

11.2.7 For the grate-fired boiler's combustion system, electrical interlocks shall be provided between the primary fan, secondary fan, main induced-draft fan, and stokehold feeding device, and grate reduction boxes. In the case of failure of the induced-draft fan, the electrical interlocks automatically cut off the primary and secondary fans and the densified biofuel supply; in the case of failure of the primary and secondary fans, they automatically cut off the densified biofuel supply.

11.2.8 Boilers shall be equipped with an over-temperature and over-pressure automatic protection device for the combustion chamber, and the automatic adjustment device for combustion chamber temperature and outlet heat medium temperature. Boilers should be equipped with a combustion process automatic adjustment device. The flue gas mixing chamber shall be provided with an automatic temperature adjustment device.

11.2.9 For the over-limit alarm for operation safety-related parameters such as pressure, water level, temperature, etc., at least two detection methods with over-limit alarm and interlock protection shall be provided, and detailed alarm records containing alarm information, alarm time, etc. shall be kept.

11.2.10 The variable frequency-based method should be adopted for boiler water level continuous adjustment, air volume adjustment, grate speed regulation, feeding adjustment, etc.

11.2.11 The steam drum shall be equipped with an automatic limit low-level and over-pressure protection device and an automatic feedwater regulation device, and should be equipped with an automatic regulation device for continuous water feeding. A standby electric feed pump shall be equipped with an automatic switch-in device.

11.2.12 Boilers shall be provided with automatic protection for the following cases:

1　The heat medium pressure at the boiler outlet exceeds the specified value.

2　The heat medium temperature at the boiler outlet exceeds the specified value.

3　The circulating pump shuts down.

11.2.13 The boiler control device shall have the function of flue gas temperature indication and over-limit alarm. For boilers with a rated evaporation capacity greater than or equal to 20 t/h or a rated thermal power greater than or equal to 14 MW, their control devices shall have a flue gas temperature recording function.

11.2.14 For the boilers with a rated evaporation capacity greater than 10 t/h or a rated thermal power greater than 7 MW, the control device shall have the function of flue gas oxygen content indication; for the boilers with a rated evaporation capacity greater than or equal to 20 t/h or a rated thermal power greater than or equal to 14 MW, the control device shall also have the function of flue gas oxygen content recording.

11.2.15 The control and alarm functions of organic fluid boilers shall meet the following requirements:

1 The boiler shall have an alarm function, and its control system should be PLC-based and shall be a self-contained system to achieve fully automatic control.

2 An oxygen analysis interface shall be reserved at the boiler tail flue duct to detect the oxygen content in the flue gas.

3 The boiler heat load shall be automatically adjusted according to the temperature change of the medium.

4 The boiler shall be provided with perfect ignition program control and furnace flameout protection.

5 Emergency buttons for boiler control and electrical equipment shall be installed at safe and easy-to-use positions.

11.2.16 The organic fluid boilers shall be equipped with automatic protection for automatic shutdown in any of the following cases:

1 The liquid level in the expansion tank drops below the limit level.

2 The boiler tapping temperature exceeds the allowable value.

3 The boiler tapping pressure exceeds the allowable value.

4 The circulating pump stops.

5 The furnace temperature exceeds the allowable value.

6 The furnace flames out.

7 The flue gas temperature exceeds the allowable value.

8 The flow rate of heat transfer fluids drops to the specified minimum value.

9 The combustion equipment fails.

11.2.17 The motor-driven equipment, water and vapor pipelines and valves, dampers, etc. of the boiler plant should be equipped with remote controls.

11.2.18 The control system of the boiler plant shall be provided with an UPS of sufficient capacity.

12 Chemical Test

12.0.1 The boiler plant shall be provided with a laboratory for water quality analysis and sampling test of heat production, moisture content, etc. of densified biofuel.

12.0.2 The boiler plant laboratory shall be able to do the following vapor and water tests:

1. For a steam boiler plant, the laboratory shall be capable of testing the suspended matter, total hardness, total alkalinity, pH, dissolved oxygen, dissolved solids, sulfate and chloride, etc. When using phosphate for water treatment inside the boiler, the laboratory shall be able to test the phosphate content.

2. For a hot-water boiler plant, the laboratory shall be capable of testing the suspended matter, total hardness, total alkalinity, and pH. When using chemicals for water treatment outside the boiler, the laboratory shall be able to test the dissolved oxygen.

13 Civil Works, Electrical and Utility System

13.1 Civil Works

13.1.1 The architectural design shall determine the building plan layout, spatial organization, architecture, color, and envelope structure according to the process requirements, taking into account the surrounding environment, natural conditions, construction materials, construction technology, etc.

13.1.2 For the boiler plant built in stages, technical measures shall be considered in civil works design.

13.1.3 Passages and openings for installation, through which the largest handling parts can pass, shall be reserved in the buildings; and the openings may be arranged at door and window openings or non-load-bearing walls.

13.1.4 Guardrails shall be set around the holes for equipment lifting and the elevated platforms.

13.1.5 Platforms and ladders shall be made of non-combustible and non-slip materials. The width of the operating platforms shall not be less than 800 mm, and the width of the ladders should not be less than 600 mm. The clear height above the platforms and ladders shall not be less than 2 m. The slope of the frequently used steel ladders should not be greater than 45°.

13.1.6 The determination of live loads on the floors, grounds, and roofs of buildings and structures shall comply with the current national standard GB 50041, *Code for Design of Boiler Plant*.

13.2 Electrical

13.2.1 The power load level and power supply mode of the boiler plant shall be determined in accordance with the current national standard GB 50052, *Code for Design Electric Power Supply Systems* based on the process requirements, boiler capacity, importance of heat load, environmental characteristics, etc.

13.2.2 The standard illuminance, color rendering index, and power density of artificial lighting for workplaces in the rooms and structures of the boiler plant shall comply with the current national standard GB 50034, *Standard for Lighting Design of Buildings*.

13.2.3 Local lighting shall be provided at the boiler water level indicators, pressure meters, instrument panels, and other locations where a higher illuminance is required. Emergency lighting should be provided at the places and passages where boiler water level indicators, pressure meters, feed pumps, and other equipment are installed and other key operations are performed. The illuminance for both local and emergency lighting should be greater than 50 lx.

13.2.4 The protection level of electrical equipment in the unloading area shall not be inferior to IP54. The dust explosion-proof grade shall comply with the current national standard GB 12476.1, *Electrical Apparatus for Use in the Presence of Combustible Dust—Part 1: General Requirements*.

13.2.5 Anti-static and earthing measures shall be taken for densified biofuel charging pipelines and equipment.

13.2.6 Earthing for lightning protection shall be provided for chimneys and outdoor equipment.

13.2.7 Communication facilities shall be set for the boiler plant.

13.3 Heating, Ventilation and Air Conditioning

13.3.1 The design of heating, ventilation and air conditioning for boiler plants shall comply with the current national standards GB 50019, *Design Code for Heating Ventilation and Air Conditioning of Industrial Buildings*; and GB 50041, *Code for Design of Boiler Plant*.

13.3.2 In severe cold regions, the densified biofuel conveying system shall consider the compensation for the heat carried away by mechanical exhaust or exhaust of the dedusting system.

13.3.3 Densified biofuel storehouses should be naturally ventilated. When mechanical ventilation is used, the ventilators and motors shall be of explosion-proof type and connected directly, with no recirculation of the indoor air. Ventilators may serve as emergency exhaust fans, and the emergency ventilation rate shall be determined on the basis of an air change rate not less than 12 times/h.

13.3.4 For the densified biofuel conveying system, closure measures and local ventilation and dedusting measures shall be taken at the transfer points and at the dry mechanical ash and slag discharge outlets to prevent dust diffusion.

13.3.5 The control room, the instrumental analysis room of the laboratory, and truck scales should be provided with air conditioning if they are used in summer.

13.4 Water Supply and Drainage

13.4.1 The boiler plant shall have reliable feedwater supply sources.

13.4.2 Densified biofuel storehouses shall be provided with water supply for sprinkling.

13.4.3 The cooling water for boilers and their auxiliaries should be recycled.

13.4.4 Drainage measures shall be arranged for the operation floor, water treatment room, pump room and ash discharge floor of the boiler plant. The equipment room in the basement shall be provided with a water sump and devices to remove the accumulated water.

14 Environmental Protection and Energy Saving

14.1 Pollution Prevention and Control

14.1.1 The air pollution control for boiler plants shall meet the following requirements:

1. The air pollutant emission of boilers shall comply with Articles 8.2.1 and 8.2.2 of this code.

2. Soot blowers shall be selected through techno-economic comparison according to such factors as the initial soot emission concentration at the outlet under the rated evaporation capacity or rated thermal power, fuel composition, soot properties, soot blower's adaptability to loads, etc.

3. Soot blowers and their ancillaries shall be provided with measures against corrosion and wear and shall be provided with reliable sealed ash discharge devices to ensure airtight conveyance and storage. The collected ash may be utilized for different purposes.

4. Sampling holes and monitoring holes shall be set for the flue gas emission system in accordance with the current standards of China GB/T 16157, *The Determination of Particulates and Sampling Methods of Gaseous Pollutants Emitted from Exhaust Gas of Stationary Source*; and HJ/T 397, *Technical Specifications for Emission Monitoring of Stationary Source*, and be provided with working platforms. For the boilers with a single-set rated evaporation capacity of 20 t/h or above or a single-set rated thermal power of 14 MW or above, automatic monitoring equipment for pollutant discharge shall be installed.

5. A negative-pressure dust collecting duct shall be arranged near the inlet of the material pit. Densified biofuel unloading areas shall be far away from the places with strict dust control requirements, and isolating, dust-control, and explosion-proof measures shall be taken.

6. For the equipment and locations that give rise to dust emissions, such as transfer points of the conveying system, dry mechanical ash and slag removing sites, etc., sealing measures shall be taken to prevent dust diffusion, and local ventilation and dust control devices shall be provided.

7. Feeding systems shall be airtight, and provided with negative-pressure dust-collecting ducts, fans, and dust collectors. The feeding and its dust-collecting system shall be explosion-proof, flame-retardant, and

explosion venting.

8 Galleries of unsealed feeding equipment may be provided with spraying devices, and ventilators may be installed to reduce dust concentration.

9 Pollutant discharge monitoring ports shall be set at the ventilation and dust control device outlets of feed bins, ash storehouses, material pits, transfer points of conveying system, dry mechanical ash and slag removing sites, charging inlets, etc. in accordance with the current sector standard HJ 953, *Technical Specification for Application and Issuance of Pollutant Permit—Boiler*.

14.1.2 Noise and vibration prevention and control for boiler plants shall meet the following requirements:

1 The noise control shall comply with the current national standard GB 3096, *Environmental Quality Standard for Noise*.

2 The noise values at the boiler plant boundary during the operation period shall comply with the current national standard GB 12348, *Emission Standard for Industrial Enterprises Noise at Boundary*.

3 Fans, pumps, grate drives, and other equipment shall be of low-noise type, and noise- and vibration-reducing measures shall be taken.

4 Mufflers shall be installed at the exhaust inlets of the blowers and the air inlets of the sound-isolating enclosures and cabins of equipment.

5 Mufflers should be installed on the steam release pipes of the steam boiler proper and temperature and pressure-reducing devices.

6 Vibration isolators shall be installed between equipment, such as fans, pumps, and their foundations. Equipment and pipelines shall be connected with flexible joints, and pipelines should be supported and fixed with flexible supports and hangers.

7 The sound transmission loss for the walls, floors, and soundproof doors and windows of non-standalone boiler plants shall not be less than 35 dB(A).

14.1.3 The water pollution control for boiler plants shall meet the following requirements:

1 Discharge of various types of wastewater shall comply with the current national standards GB 8978, *Integrated Wastewater Discharge Standard*; and GB 3838, *Environmental Quality Standards for Surface Water*.

2   The discharged wastewater shall be treated by water quality and quantity according to different discharge requirements, and recycled properly.

3   Boiler effluent shall be cooled to below 40 °C and then discharged.

14.1.4 Solid wastes from boiler plants shall not cause secondary pollution to the surrounding environment.

14.2 Greening

14.2.1 The boiler plant area shall be greened. The green space ratio for district boiler plants should be 20 %, and the green area of non-district boiler plants shall be planned in the overall design.

14.2.2 The green isolation zone should be set around densified biofuel storehouses and dry slag rooms.

14.3 Environmental Management and Monitoring

14.3.1 The operator of the boiler plant shall establish an environmental monitoring system and formulate a monitoring scheme to monitor the boiler's air pollutant discharge and the impact on the surrounding environment quality, save the original monitoring records, and publish the monitoring results.

14.3.2 Permanent sampling ports, sampling and testing platforms, sewage outlets, and their signs shall be designed, built, and maintained for boiler plants.

14.3.3 Sampling of exhaust gas from boilers shall be carried out at the specified pollutant discharge monitoring location according to the types of pollutants to be monitored. Where waste gas treatment facilities are available, monitoring shall be carried out behind the facilities. Monitoring and sampling of air pollutants from chimneys shall comply with the current standards of China GB 5468, *Measurement Method of Smoke and Dust Emission from Boilers*; GB/T 16157, *The Determination of Particulates and Sampling Methods of Gaseous Pollutants Emitted from Exhaust Gas of Stationary Source*; and HJ/T 397, *Technical Specifications for Emission Monitoring of Stationary Source*.

14.3.4 The automatic monitoring equipment installed for pollutant discharge shall be connected to the monitoring center of the local environmental protection department.

14.3.5 The quality control for monitoring of air pollutants from boilers shall be in accordance with the current sector standard HJ/T 373, *Technical Specifications of Quality Assurance and Quality Control for Monitoring of Stationary Pollution Source (on Trial)*.

14.4 Energy Saving

14.4.1 After a boiler plant is put into normal operation, a thermal efficiency test shall be conducted for the boiler. The test shall be carried out by a qualified organization in accordance with the current national standard GB/T 10180, *Thermal Performance Test Code for Industrial Boilers*.

14.4.2 Except for short-time loads, the operating load of the boiler should not be lower than 70 % of the rated evaporation capacity or 70 % of the rated heating capacity.

14.4.3 Boiler energy efficiency monitoring items shall cover exhaust gas temperature, excess air coefficient at the exhaust point, carbon content of slag, furnace surface temperature, etc. Boiler energy efficiency monitoring methods and performance indicators shall be in accordance with the current national standard GB/T 15317, *Monitoring and Testing for Energy Saving of Coal Fired Industrial Boilers*.

14.4.4 The boiler steam system shall be prevented from leakage, and the heat of secondary steam, condensate water, and continuous sewage shall be utilized. Measures shall be taken to achieve a recovery rate of greater than 80 % for recoverable condensate water.

14.4.5 For boiler drums, induced-draft fans, circulating water pumps and make-up water pumps of the heat supply network, etc., whether to employ a speed regulating device should be determined according to the system operation and heating load.

14.4.6 The cooling water for grate shafts, induced-draft fan bearings, circulating water pumps of the heat supply network and chemical analysis, and the waste water from slag flushing shall be recycled; the utilization rate of recycled water shall be greater than 80 %. Water-saving equipment and apparatus shall be selected.

14.4.7 The boiler plant shall be equipped with low-voltage capacitors for reactive power compensation, to make the power factor greater than 0.9 at the maximum electrical load. The 10 kV high-voltage system shall be provided with dedicated metering cabinets, the low-voltage system shall be equipped with incoming line cabinets, and the lighting line shall be equipped with separate meters.

15 Fire Protection

15.0.1 The classification of fire hazards and fire endurance rating of boiler plant buildings and structures shall be in accordance with Table 15.0.1.

Table 15.0.1 Classification of fire hazards and fire endurance rating of boiler plant buildings and structures

S/N	Building and structure	Classification of fire hazards	Fire endurance rating
1	Boiler house	D	II
2	Control room	D	II
3	Office, duty room	D	II
4	Locker room, shower room	E	II
5	Fan room	D	II
6	Dedusting structure	D	II
7	Chimney	D	II
8	Pump room, water tank room	E	II
9	Water treatment room	E	III
10	Laboratory	D	II
11	Transformer room	C	II
12	Power distribution room	D	II
13	Outdoor switchgear installation	C	II
14	Densified biofuel storehouse	C	II
15	Conveying trestle	C	II
16	Bottom-dumping bunker	C	II
17	Truck unloading trough	C	II
18	Sewage treatment structure	E	II

15.0.2 Interior decoration of various buildings in the boiler plant shall comply with the current national standard GB 50222, *Code for Fire Prevention in Design of Interior Decoration of Buildings*.

15.0.3 The spacing between densified biofuel storehouse, silo, conveying trestle, and boiler house, the spacing between those and other buildings and structures, and the fire lanes shall meet the requirements of the current national standard GB 50016, *Code for Fire Protection Design of Buildings*.

15.0.4 The fire water for the boiler plant may be sourced from the production and living water supply system.

15.0.5 The fire water consumption and fire hydrants in the boiler plant shall comply with the current national standard GB 50016, *Code for Fire Protection Design of Buildings*; the fire extinguisher distribution shall comply with the current national standard GB 50140, *Code for Design of Extinguisher Distribution in Buildings*.

15.0.6 Organic fluid boiler systems shall take the following fire-extinguishing measures:

1   Organic fluid boilers shall be provided with a fire extinguishing system which should use steam or nitrogen as extinguishing gas. Steam shall be used when a stable steam supply is available.

2   Gas supply for fire extinguishing shall be sufficient to fill up the furnace for at least three times within 15 min.

3   Nitrogen fire extinguishing cylinder group or nitrogen tank shall be connected to the fire extinguishing gas interface of an organic fluid boiler and shall ensure the continuous supply of nitrogen in case of fire.

15.0.7 Boiler houses shall be provided with fire detectors and automatic fire alarm devices. The selection and setting of fire detectors, and fire control equipment and their functions shall comply with the current national standard GB 50116, *Code for Design of Automatic Fire Alarm System*. The boiler houses set in civil buildings shall be equipped with sprinkler systems.

15.0.8 The control room should be equipped with carbon dioxide fire extinguishers, mobile water mist fire extinguishers, and portable water-based fire extinguishers, and they should be used jointly.

15.0.9 Indoor fire hydrants should be set in densified biofuel storehouses and boiler houses, at conveying trestles, etc., and fire compartment facilities with drencher sprinkler systems shall be provided at their connections.

15.0.10 Densified biofuel storehouses shall be equipped with an automatic fire extinguishing system, which shall comply with the current national standard GB 50016, *Code for Fire Protection Design of Buildings*.

15.0.11 Silos shall be provided with a combustible gas alarm system and an automatic fire alarm system; the upper, middle and lower parts of a silo shall be provided with a fire extinguishing system respectively, and the silo top shall be installed with sprinklers.

15.0.12 The structural members of conveying trestles shall be made of non-combustible materials. When a conveying trestle adopts a steel structure, fire protection measures shall be taken accordingly.

15.0.13 Measures to prevent backfire shall be taken for the boiler fuel feeding system.

15.0.14 The dust-removal ducts and components in densified biofuel conveying system shall be made of non-combustible materials.

16 Occupational Health and Safety

16.1 Occupational Safety

16.1.1 Proper fire separation distance, safety clearance and convenient exits for fire protection and safety evacuation shall be designed for boiler houses, auxiliary rooms, dumping devices, densified biofuel storehouses, silos, biofuel conveying equipment, dry slag rooms, dry ash storage bins, rest quarters, flammable and explosive hazardous places, underground structures, etc. The safety evacuation facilities shall have adequate lighting and conspicuous evacuation signs.

16.1.2 The lightning and explosion protection design for boiler plants shall comply with the current national standards GB 50057, *Code for Design Protection of Structures Against Lightning*; and GB 50058, *Code for Design of Electrical Installations in Explosive Atmospheres*.

16.1.3 The design of pressure vessels and pressure piping in boiler plants shall comply the current standards of China GB 150, *Pressure Vessels*; TSG G0001, *Boiler Safety Technical Supervision Administration Regulation*; TSG 21, *Supervision Regulation on Safety Technology for Stationary Pressure Vessel*; GB/T 20801, *Pressure Piping Code—Industrial Piping*; and TSG D0001, *Pressure Pipe Safety Technology Supervision Regulation for Industrial Pressure Pipe*.

16.1.4 For the machines with exposed transmissions and rotating components or the equipment with protruding edges, angles, and projecting parts that people might reach, reliable protective devices, safety signs or safe operating areas shall be set.

16.1.5 When the air inlet or air duct of the fan leads to the open air, a protective net shall be installed or other safety measures shall be taken.

16.1.6 Straight ladders at chimneys, person-accessible roofs, etc. shall be provided with safety cages.

16.1.7 The balconies, verandas, indoor ambulatories, interior courtyards, outdoor stairways, person-accessible roofs, platforms, floor openings, etc. of boiler plants shall be provided with fixed guardrails or parapets. Slip prevention measures shall be taken for stairs, steel ladders, and platforms.

16.1.8 For the design of roads in boiler plants, the personnel, vehicles, and material transport shall be arranged reasonably, and warning signs shall be set at dangerous locations to prevent traffic accidents.

16.1.9 Conspicuous safety signs and safety colors shall be provided in boiler plants according to equipment and facility functions, safety requirements, protection and warning needs, fire protection regulations and operating environments, etc., considering the plan layout, process setup, equipment requirements, transportation, etc. The design of safety signs and safety colors shall comply with the current national standards GB 2894, *Safety Signs and Guideline for the Use*, and GB 2893, *Safety Colours*.

16.2 Occupational Health

16.2.1 Boiler plants shall be provided with dust prevention measures. Comprehensive measures such as airtight operation, hydraulic cleaning and dedusting shall be taken for the locations of unloading, storage, conveying and feeding, and boiler and ash removal systems; dust concentration in indoor air shall comply with the current national regulations on occupational exposure limits for hazardous agents in the workplace.

16.2.2 For the design of boiler plant noise and vibration control, noise and vibration shall first be controlled at sources by means of muffling, sound insulation, sound absorption, vibration isolation, and vibration damping measures, and shall comply with the current national standards GBZ 1, *Hygienic Standards for the Design of Industrial Enterprises*; GBZ 2.2, *Occupational Exposure Limits for Hazardous Agents in the Workplace—Part 2: Physical Agents*; GB/T 50087, *Code for Design of Noise Control of Industrial Enterprises*; and GB 50463, *Code for Design of Vibration Isolation*.

16.2.3 The indoor air design parameters of each workplace in the boiler plant shall comply with the current national standards GB 50041, *Code for Design of Boiler Plant*; and GB 50019, *Design Code for Heating Ventilation and Air Conditioning of Industrial Buildings*.

17 Operation and Maintenance

17.1 Boiler Operation

17.1.1 To reduce ash deposits and ash removal frequency, the following measures should be taken for boilers:

1. For grate-fired steam boilers, both sides of boiler tube rows should be set with ash removal gates to facilitate removing the ash deposits in flue tubes and water pipes.

2. For circulating fluidized bed boilers, ash removal gates should be set at such parts with great flow resistance as economizers and air preheaters of vertical flue ducts, etc.

3. Boiler heat exchange tube spacing and resistance for flue duct diameter should be considered. The flow cross-sectional area should be as large as possible to reduce ash deposition provided that the flow rate requirements are met.

4. Soot blowers should be provided at such heating surfaces as superheater, convection pass, economizer, and air preheater to prevent ash deposition and reduce the frequency of manual ash removal.

17.1.2 The boiler furnace wall and flue gas and air ducts shall have good sealing and thermal insulation performance. The outside surface temperature of the furnace body shall comply with the current sector standard NB/T 47034, *Specification for Industrial Boilers*.

17.2 Maintenance

17.2.1 Maintenance room or yard may be provided in the boiler plant for maintenance and minor repairs.

17.2.2 For double-layered boiler plants and for single-layered boiler plants with a single steam boiler having a rated evaporation capacity greater than or equal to 10 t/h or with a single hot water boiler having a rated thermal power greater than or equal to 7 MW, a hoist shall be provided above the boiler to lift objects from the ground to the top of the boiler. A lifting hole shall be set for interfloor lifting where needed.

17.2.3 Hoists for maintenance should be set above the auxiliary equipment, including fans, water pumps, heat exchangers, thermal deaerators, and ion exchangers with barrel flanges, or spaces should be reserved for lifting operation.

17.3 Emergency Response Plan

17.3.1 The operator of the boiler plant shall develop a special emergency response plan in accordance with the relevant procedures and requirements, to specify the rescue procedure and specific emergency rescue measures.

17.3.2 In the event of heat supply interruption, the operator shall initiate the emergency shutdown procedure and start emergency safety measures such as the boiler parameter overlimit alarm, the automatic overtemperature and overpressure protection, and the electrical interlock to automatically cut off the densified biofuel supply.

Explanation of Wording in This Code

1 Words used for different degrees of strictness are explained as follows in order to mark the differences in executing the requirements in this code.

   1) Words denoting a very strict or mandatory requirement:

      "Must" is used for affirmation; "must not" for negation.

   2) Words denoting a strict requirement under normal conditions:

      "Shall" is used for affirmation; "shall not" for negation.

   3) Words denoting a permission of a slight choice or an indication of the most suitable choice when conditions permit:

      "Should" is used for affirmation; "should not" for negation.

   4) "May" is used to express the option available, sometimes with the conditional permit.

2 "Shall meet the requirements of…" or "shall comply with…" is used in this code to indicate that it is necessary to comply with the requirements stipulated in other relative standards and codes.

List of Quoted Standards

GB 150,	*Pressure Vessels*
GB 1576,	*Water Quality for Industrial Boilers*
GB 2893,	*Safety Colours*
GB 2894,	*Safety Signs and Guideline for the Use*
GB 3096,	*Environmental Quality Standard for Noise*
GB 3838,	*Environmental Quality Standards for Surface Water*
GB 5468,	*Measurement Method of Smoke and Dust Emission from Boilers*
GB 8978,	*Integrated Wastewater Discharge Standard*
GB 12348,	*Emission Standard for Industrial Enterprises Noise at Boundary*
GB 12476.1,	*Electrical Apparatus for Use in the Presence of Combustible Dust—Part 1: General Requirements*
GB 13271,	*Emission Standard of Air Pollutants for Boiler*
GB 50016,	*Code for Fire Protection Design of Buildings*
GB 50019,	*Design Code for Heating Ventilation and Air Conditioning of Industrial Buildings*
GB 50034,	*Standard for Lighting Design of Buildings*
GB 50041,	*Code for Design of Boiler Plant*
GB 50052,	*Code for Design Electric Power Supply Systems*
GB 50057,	*Code for Design Protection of Structures Against Lightning*
GB 50058,	*Code for Design of Electrical Installations in Explosive Atmospheres*
GB 50116,	*Code for Design of Automatic Fire Alarm System*
GB 50140,	*Code for Design of Extinguisher Distribution in Buildings*
GB 50187,	*Code for Design of General Layout of Industrial Enterprises*
GB 50222,	*Code for Fire Prevention in Design of Interior Decoration of Buildings*
GB 50463,	*Code for Design of Vibration Isolation*
GBZ 1,	*Hygienic Standards for the Design of Industrial Enterprises*
GBZ 2.2,	*Occupational Exposure Limits for Hazardous Agents in the*

Workplace—Part 2: Physical Agents

GB/T 10180, *Thermal Performance Test Code for Industrial Boilers*

GB/T 15317, *Monitoring and Testing for Energy Saving of Coal Fired Industrial Boilers*

GB/T 16157, *The Determination of Particulates and Sampling Methods of Gaseous Pollutants Emitted from Exhaust Gas of Stationary Source*

GB/T 20801, *Pressure Piping Code—Industrial Piping*

GB/T 50087, *Code for Design of Noise Control of Industrial Enterprises*

HJ 75, *Specifications for Continuous Emissions Monitoring of SO_2, NO_X, and Particulate Matter in the Flue Gas Emitted from Stationary Sources*

HJ/T 373, *Technical Specifications of Quality Assurance and Quality Control for Monitoring of Stationary Pollution Source (on Trial)*

HJ/T 397, *Technical Specifications for Emission Monitoring of Stationary Source*

HJ 953, *Technical Specification for Application and Issuance of Pollutant Permit—Boiler*

NB/T 47034, *Specification for Industrial Boilers*

NB/T 47051, *Technical Specification for Industrial Boiler Control Devices*

NB/T 47062, *Biomass Molded Fuel Fired Boilers*

SY/T 0524, *Code for Heater System with Heat-Transfer Oil*

TSG 21, *Supervision Regulation on Safety Technology for Stationary Pressure Vessel*

TSG D0001, *Pressure Pipe Safety Technology Supervision Regulation for Industrial Pressure Pipe*

TSG G0001, *Boiler Safety Technical Supervision Administration Regulation*

TSG G0002, *Supervision Administration Regulation on Energy Conservation Technology for Boiler*